YOUR KNOWLEDGE HAS VALUE

- We will publish your bachelor's and
  master's thesis, essays and papers

- Your own eBook and book -
  sold worldwide in all relevant shops

- Earn money with each sale

Upload your text at www.GRIN.com
and publish for free

Bibliographic information published by the German National Library:

The German National Library lists this publication in the National Bibliography; detailed bibliographic data are available on the Internet at http://dnb.dnb.de .

Imprint:

Copyright © 2016 GRIN Verlag, Open Publishing GmbH
Print and binding: Books on Demand GmbH, Norderstedt Germany
ISBN: 978-3-668-18973-7

This book at GRIN:

http://www.grin.com/en/e-book/318703/examplary-failure-modes-and-effects-analysis-fmea-of-a-flashlight

Paul Scholz

Examplary Failure Modes and Effects Analysis (FMEA) of a Flashlight

GRIN Publishing

GRIN - Your knowledge has value

Since its foundation in 1998, GRIN has specialized in publishing academic texts by students, college teachers and other academics as e-book and printed book. The website www.grin.com is an ideal platform for presenting term papers, final papers, scientific essays, dissertations and specialist books.

Visit us on the internet:

http://www.grin.com/

http://www.facebook.com/grincom

http://www.twitter.com/grin_com

1. Table of Contents

2. Table of Figures

3. Introduction of FMEA

a. History

FMEA was developed in the 1960 during a NASA project and formerly mainly used in cutting edge aerospace projects in order to avoid quality problems to occur. The method was standardized in Germany in 1980 named "Failure Effect Analysis" originally. Since it first was not used in production engineering, its deployment and modification especially for this industry sectors has been triggered by the introduction of quality system guideline by Ford during the late 1980s. The original method was extended in the sense of improving the needs of the automotive industry and its scope changed to a system FMEA for products and processes. The main modification appeared to be the introduction of risk priority number (RPN). The RPN provides an indication of the urgency of the various failure modes to the developing company. Thus it can provide guidance when searching for improvement. Three indicators mainly determine the value of the RPN: likelihood of occurrence, significance of the effects of the failure, likelihood that it will be detected.

b. General Facts and Benefits

FMEA is the short term for "Failure Mode and Effects Analysis". According to ISO 9004-1 "risk assessment should be undertaken to assess the potential for, and the effect of, possible failures or faults in products or processes" and FMEA is considered to be a suitable method to this purpose. The benefit of a FMEA is to identify reveal potential faults during the development phase of products or new production methods already. Appropriate action to avoid failures to occur can already be implemented at the planning stage. Therefore, FMEA uncovers the correlations between faults and factors that influence quality in a systematic manner. The knowledge about these correlations can be very valuable for companies and should be made available within the company in order to take quality harming aspects into account as early as possible. Adjustments to hinder quality problems and other failures from occurring can save a lot of money and also customers' credit to the company. A sustainable and sophisticated FMEA system provides a source of experience acquired from previous solutions - empirical knowledge which in other ways hardly can be kept over times. Originally FEMA intended to "record and systematically evaluate information relating to the reliability, safety and maintenance of a system through inductive analysis of types of failures sustained by all of the components and of their effects". All effort going along with the maxim that a minimum number of possible of failure modes leads to a faultless product. Further benefits of FMEA are:

- Reduces process development time and costs

- Documents and tracks risk reduction activities

- Helps to identify critical-to-quality characteristics

- Helps increase customer satisfaction and safety

The method of FMEA is closely related to quantitative methods such as Fault-Tree Analysis and incident-sequence analysis.

c. Types

Various types of FMEAs can be found like e. g. Design FMEA, Process FMEA and System FMEA. Differences between the mainly origin form associated objective during the planning phase, when the FMEA is drawn up. The System FMEA is conducted concerning the work schedule of manufacturing – revealing the potential failures inherent in the design from the point of production.

d. Method Description

1) Organizational Preparation for FMEA

The first step is to define the parts or processes which are needed for the FMEA. After the team is defined, responsible persons are assigned to the team. Furthermore, a project schedule is implemented.

2) Preparing the subject of FMEA

In this step, the systematically structure of the task needs to be described in a detailed way to ensure a smooth process. The FMEA task are assigned to the team members.

3) Conducting the analysis

The conducted analysis contains potential faults, causes and effects of faults and potential risks.

4) Evaluating the analysis results

In this step, measures are identified to eliminate the weaknesses of the product. Applied methods are design modifications, detection methods and damage limitation methods.

5) Following up deadlines and checking effectiveness

The final step of FMEA is to make the measures effective and control that the tasks are done before the deadlines. If it is necessary to conduct a new risk assessment, new measures need to be planned.

4. Preparing the Content for FMEA

a. Product Structure

Before the final FMEA sheet can be filled out the required data needs to be prepared. In our case, we need to organize the components of the flashlight into a graphic to get a structured overview of the components. The more detailed the preparation steps are done the easier it is in the end to put the values into the FMEA form. A convenient way to do this is to use a tree graphic to divide the components using several structural levels. To be able to conduct a reliable FMEA analysis all single components need to be listed. By disassembling the flashlight, the single parts could be described and analysed:

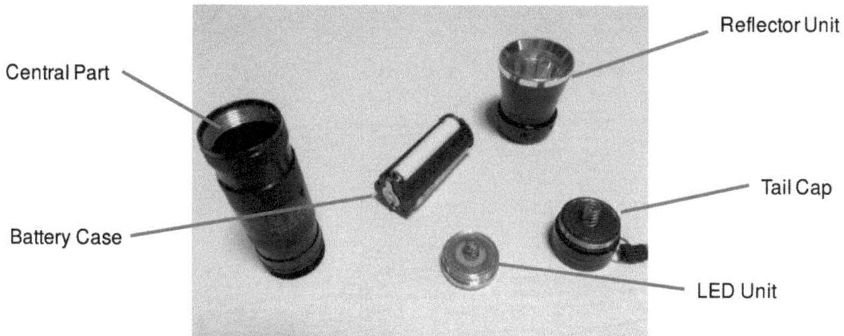

Figure 1: Flashlight Components of Sublevel 1

The components of the flashlight can be structured into 3 sublevels. Above the components of sublevel 1 are shown which namely are Central Part, Battery Case, Reflector Unit, Tail Cap and LED Unit. The idea is to get more and more specific with each additional sublevel. The last sublevel of the structural tree should represent all single, not dividable components. For example, the Reflector Unit can be further divided into Reflector and Lens. The Tail Cap is assembled out of a Switch, Electrical Spring and O- Ring. The Battery Case and 3xAA 1.5V Batteries represent sublevel 3 of the Energy Unit. The LED Unit can be divided into LED, Electrical Spring and Solder Joint. Since the Central Part is just a shell with threads on both sides, it can not be further divided.

b. Structural Tree

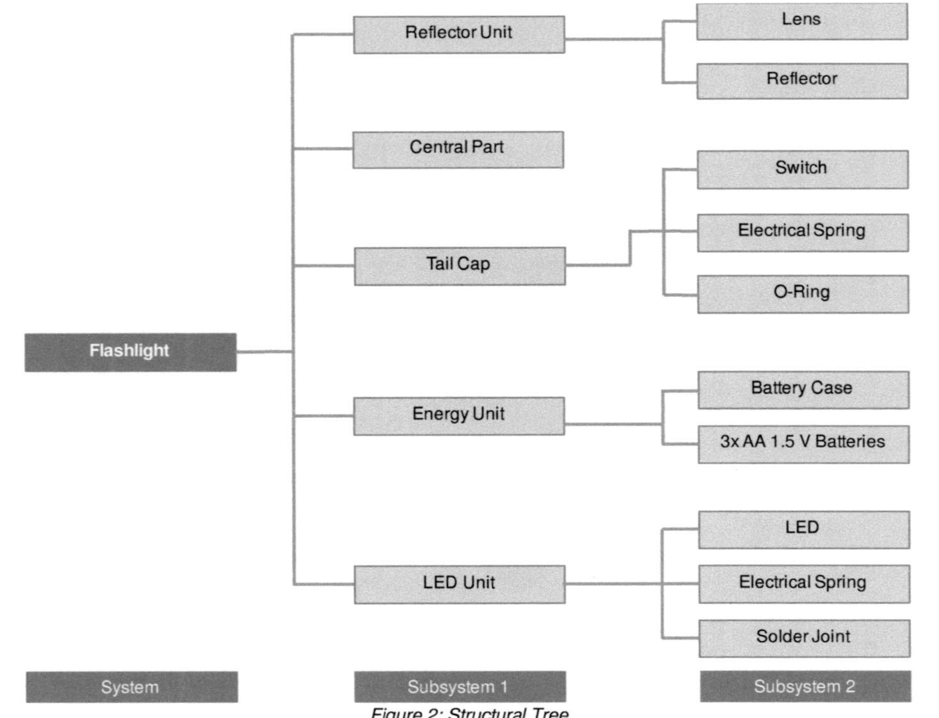

Figure 2: Structural Tree

c. Functions

For all the determined elements of the systems and the subsystems several functions are defined. For this reason, the structural tree is extended to the functional tree. In the following chapter a short description of each function and sub function shall be given.

Main function

- The *flash light* has to deliver a light beam which has to be light enough to bright dark spots accurately. The brightness of the beam should be stable and constant over time.

Functions of Sublevel 1

- The *reflector part* has to protect the parts it contains which are the *lens* and the *reflector* itself. Furthermore, it has to ensure a stable connection to the *middle part.*

- The *middle part* has the function to connect the *reflector part* and the *tail cap.* Furthermore, it has to protect the *energy unit* which is located in its interior.

- The *tail cap* has to enclose its inner parts. These inner parts are the *O-ring* and the *switch.* Its main function is the connection of the push mechanics through the *switch* use and the corresponding electrical components.

- The *energy unit* has to provide the energy of the *batteries* it stores.

- The *LED unit* fixates the components it contains. The main function is the emission of the light beam delivered by the *LED.*

Functions of Sublevel 2

- The *lens* has to protect the *LED.* Furthermore, the light beam should be able to pass the *LED.*

- The *reflector* has to ensure that the light beam is stable through its mirrored surface. Additionally, it also protects the *LED.*

- The *O-ring* has to protect the sensitive electrical components and mechanics against environmental influences.

- The *switch* has to open or close the electrical circuit.

- The *spring of the tail cap* has to ensure the electrical connection between the *energy unit* and the *tail cap*.

- The *battery case* guarantees a stable fixation of the *batteries*.

- The *batteries* themselves have to provide neither higher nor lower as 1.5V each.

- The *LED* has to transfer electricity in to light.

- The *spring of the LED* has to ensure the electrical connection between the *energy unit* and the *LED*.

- The *solder joint* should work as a mechanical fixation of the *LED unit*.

d. Functional Tree

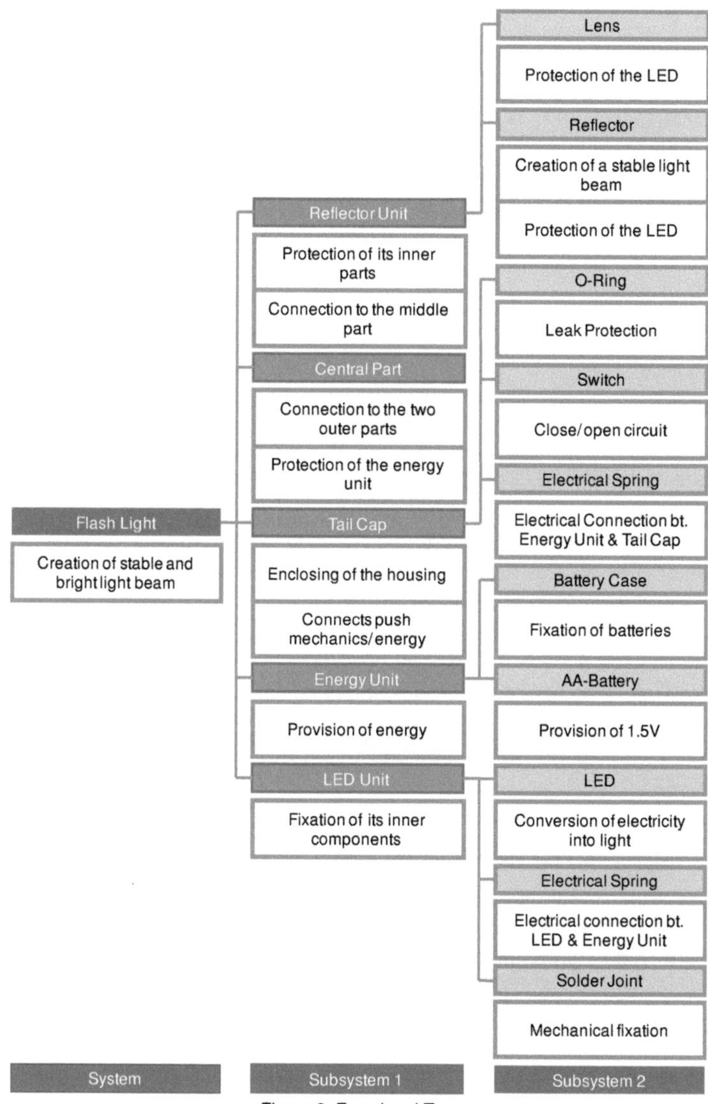

Figure 3: Functional Tree

e. Possible Failures

Flashlight

For system Flashlight 3 errors may occur. First of all, it is possible that the lamp does not light. Also some malfunction can occur so that the lamp light is weak. The 3rd possible failure is that the lamp light flickers.

Reflector Unit

Since the function of the Reflector Part is to bundle the light rays, a failure mode which can occur is that light rays are bundled badly.

Lens

The Lens can be dull. This may occur through damage or be a production failure. Also condensed water can lead to a dull Lens.

Reflector

The failure of the Reflector can be a broken or dull surface. Damage, production failure or condensed failure can be the reason for the failure.

Central Part

A leakage in the Middle Part is a failure mode. Fluids or dust can enter and lead to damage inside the Middle Part.

Tail Cap

The Tail Cap is the cap of the Flashlight and carries Switch, sealing and a spring. A problem occurs when there is no electricity provided or the circuit is not closed. If the Tail Cap doesn't work properly the Flashlight will not work.

O-Ring

The O-Ring protects the inside of the Flashlight from water and other fluids. In case of a leakage the Flashlight might break.

Switch

Failure modes of the switch are open circuit so that no electricity is supplied and loose contact so that energy is supplied discontinuously.

Spring of Tail Cap

If the spring is broken there won't be energy supply. Also the spring can be too short and thus energy supply will be intermittently and influenced by vibration.

Energy Unit

Common failure modes for the Energy Unit are when there is weak or no energy supply. Also an energy supply which is not steadily is a problem. This leads to flickering lamp light.

Battery Case

There are some possible malfunctions with the Battery Case. First of all there might be wrong installations of Battery Case. It is possible to mix up plus and minus pole. The Battery Case also might be worn out so that the Battery doesn't fit or contacts don't work properly. The contact between Battery Case and Battery can be damaged and avoid current flow.

AA-Battery

Like the Battery Case, the AA-Battery can be mixed up and installed in the wrong way. A failure is also an empty or low Battery.

LED Unit

The LED unit can have several failures. It can have an error which leads to a wrong function of the LED or it cannot work at all. Furthermore, it is a problem when the LED supplies light inconsistently.

LED

Electrical contacts can be torn off after a long time and lead to other malfunctions. Short circuit and weak contacts lead to electrical malfunctions. A dull LED is a malfunction which can be caused by production failure.

LED Unit Spring

The LED Unit Spring has the same malfunctions like the spring of the Tail Cap. It can be too short or broken.

Solder Joint

Failure modes for the Solder Joint can be broken wire which lead to no current flow. In addition, a bad Soldering can have no contact between solder parts and thus no electricity flow as well. Lastly, the material can have a failure. For example, the material can be too brittle and break.

f. Malfunction Tree

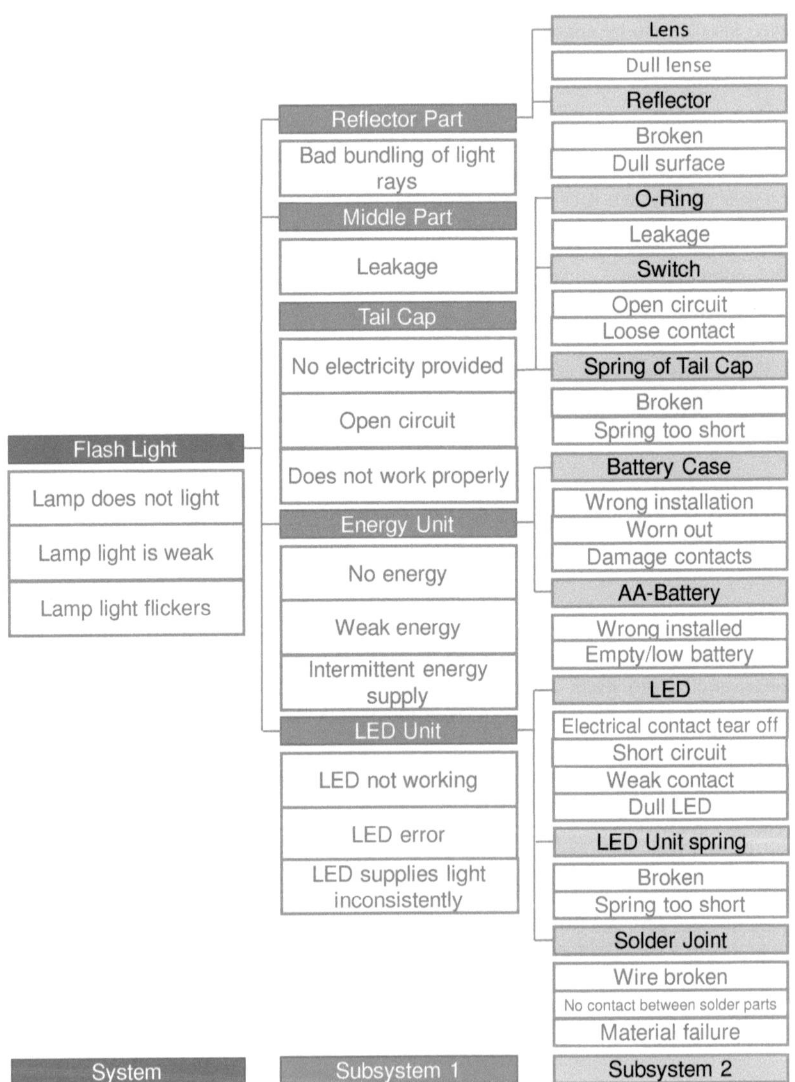

Figure 4: Malfunction Tree

g. Failure Trees

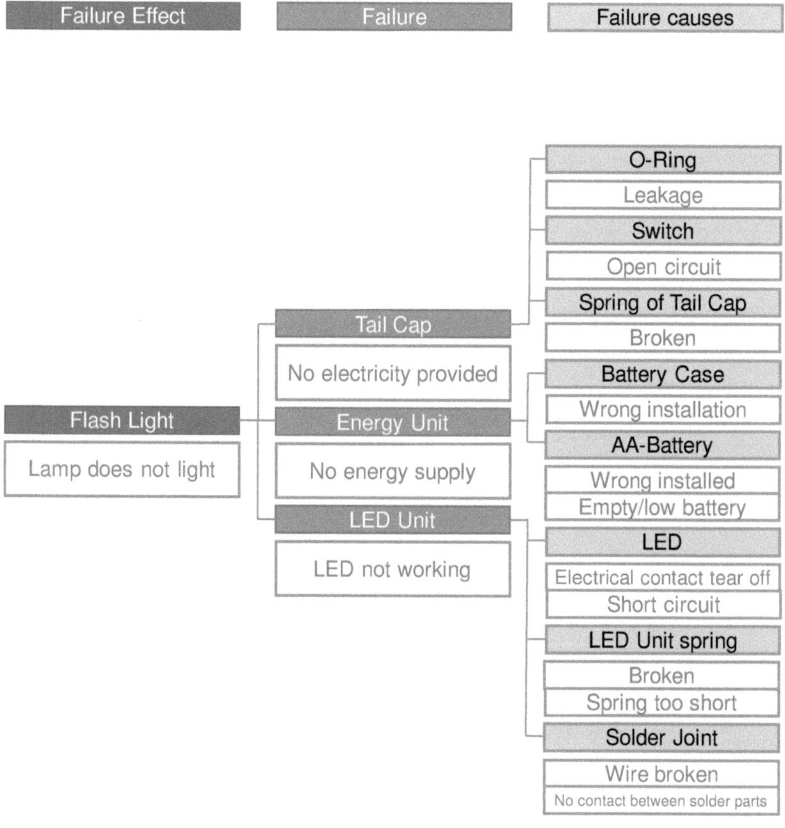

Figure 5: Failure Tree 1

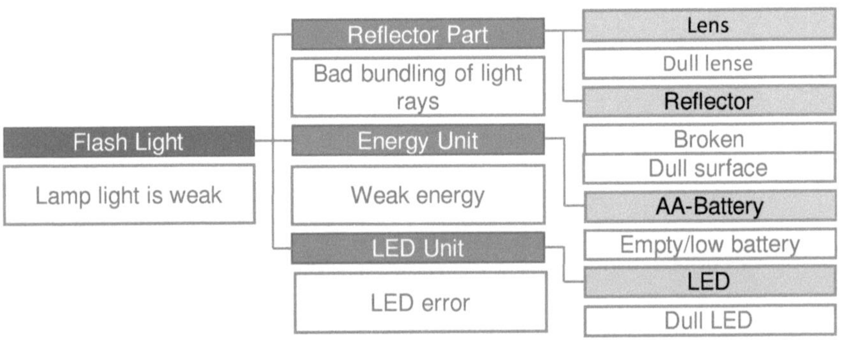

Figure 6: Failure Tree 2

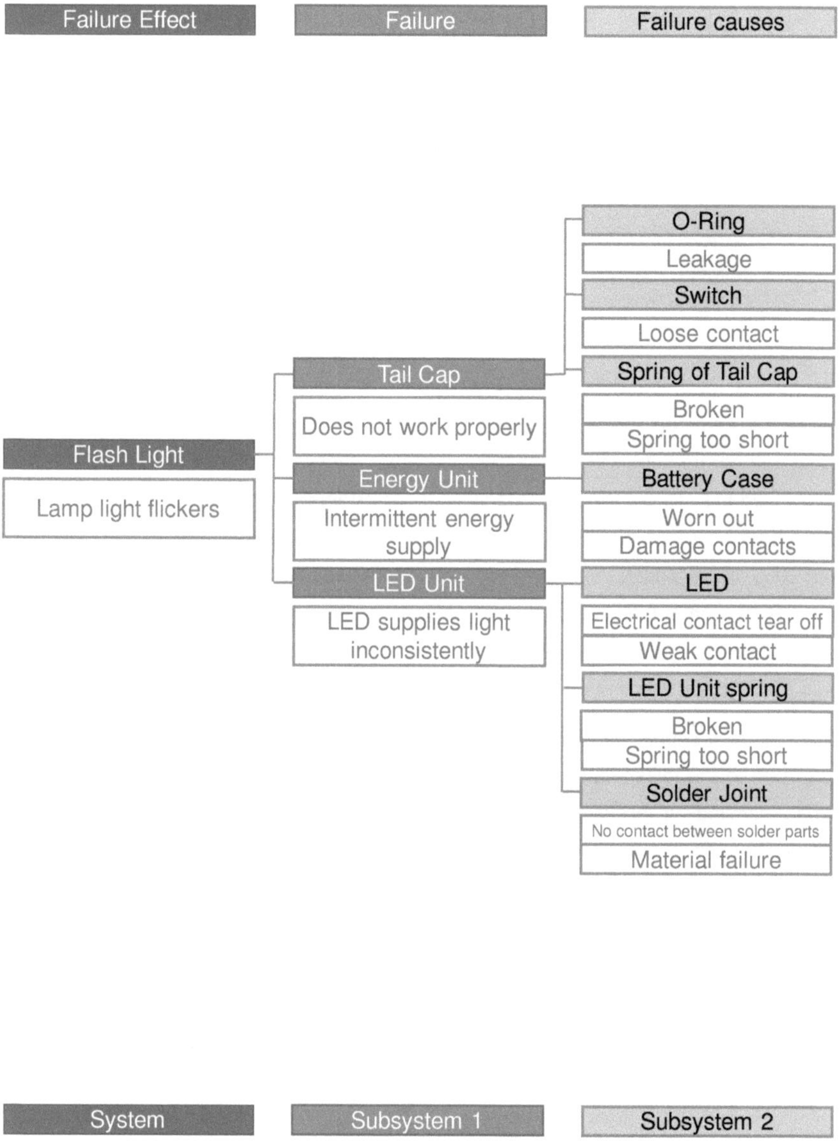

Figure 7: Failure Tree 3

5. FMEA Form

In the final step the possible faults, the possible effect of these faults and the causes are inserted into the FMEA form for each function. These data are obtained from the possible failure trees, explained in the previous chapter. The flashlight in our example has three possible effects. The possible effect "Flashlight does not light" has three possible failures and totally twelve possible causes. The effect "Light of flashlight is weak" has also three possible failures but only five possible causes. The last effect "Light of flashlight flickers" has three possible failures with twelve possible causes. Furthermore, for each combination of "Fault, Effect and Cause" a remedial measure and a detection measure is determined. Then it is defined how severe each failure effect is, how likely each failure cause is to come up and how likely it is that the failure cause will be detected. Therefore, a risk number from the range 1 to 10 is allocated, where "1" denotes a very low risk and "10" a high risk. To get the Risk Priority Number (RPN) the three allocated risk numbers of each failure mode must be multiplied. The RPN is between 1, representing low risk, and 1000, representing high risk. In industry often fixed limits for RPN values are used, e.g. a RPN greater than 125 should be avoided. Additionally, one responsible person for each failure cause has to be determined, who is responsible for to implement measures until a fixed date.

The approach should be exemplary shown for the possible effect "Flashlight does not light" due to the failure "Energy unit doesn´t provide energy" because "Battery case is wrong installed". If the flashlight doesn´t work, the main function of the flashlight isn´t fulfilled. In such a case the customer will be highly upset and therefore the R_nS value is set to 7 out of 10. The probability for the battery case to be wrong installed is considered to be relatively low. To detect a wrong installed battery case an ampere and voltmeter is needed. The failure will be detected in most cases but only by using further equipment therefore the detection grade is set to 7 out of 10 and the occurrence value is set to 3 out of 10.

Failure Mode and Effects Analysis

X	System FMEA Product				System FMEA Process				

Type/model/manufacture/batch:					Subject No.:		Responsibility:		
							Sebastian, Andi		
Flashlight					Status:		Company:		
							Tsinghua University		
System no./System element:					Subject No.:		Responsibility:		
Flashlight							**Sebastian, Andi**		
Function/Task:					Status:		Company:		
Provide a focused beam of light							**Tsinghua University**		

No.	Possible Effect	R_nS	Poss. Failure	Poss. Causes	Remedial Measures	R_nO	Detection Measures	R_nD	RPN
1	Flashlight does not light	7	Tail cap doesn´t provide energy	Leakage due to broken O-Ring	Use thicker O-Ring and remove sharp edges	6	Hydrostatic test for detecting leakages	4	168
2	"	7	"	Switch doesn´t close electrical circuit	Use better and reliable switch	2	Use of ohmmeter to check electrical circuit	3	42
3	"	7	"	Spring of Tail Cap is broken	Use better and reliable spring	4	Control Spring	2	56

4	"	7	Energy Unit doesn't provide energy	Battery case is wrong installed	Apply Poka Yoke	3	Use of amperemeter and voltmeter to check polarity	7	147
5	"	7	"	AA-Battery is wrong installed	Mark right placement of battery	3	Use of amperemeter and voltmeter to check polarity	7	147
6	"	7	"	AA-Battery is empty or low	Use better and more reliable or certified batteries	6	Use of amperemeter and voltmeter to test batteries	9	378
7	"	7	LED Unit doesn't work	Electrical contact of LED torn off	Use better and reliable LED	8	Use of ohmmeter to check electrical circuit	2	112
8	"	7	"	LED is short circuited	Use LED which can work in a higher ampere range	9	Use of ohmmeter to check electrical circuit	2	126
9	"	7	"	LED Unit spring is broken	Use better and reliable spring	4	Control Spring	3	84

10		"	7	"	LED Unit spring is to short/weak	Use better and reliable spring	6	Control Spring	4	168
11		"	7	"	Wires of solder joint are broken	Use more flexible and thicker wires	5	Use of ohmmeter to check electrical circuit	3	105
12		"	7	"	No contact between solder parts	Use better solder / alloy	5	Use of ohmmeter to check electrical circuit	2	70
13	Light of flashlight is weak	Reflector Part bundles light rays bad	5	Lens is dull		Use a more polished lens / protect surface by coating	1	Use of measurement instruments to control light density	9	45
14		"	5	"	Reflector is broken	Use more shock resistance material	1	Use of measurement instruments to control light density	5	25
15		"	5	"	Reflector has dull surface	Use a more polished reflector / protect surface by coating	1	Use of measurement instruments to control light density	6	30

16	"	5	Energy unit provides weak energy	AA-Battery is empty or low	Use better and more reliable or certified batteries	6	Use of amperemeter and voltmeter to test batteries	9	270
17	"	5	LED Unit has a error	LED is dull	Protect surface of LED by coating	3	Use of measurement instruments to control light density	3	45
18	Light of flashlight flickers	6	Tail cap doesn´t work properly	Leakage due of broken O-Ring	Use thicker O-Ring and remove sharp edges	7	Hydrostatic test for detecting leakages	4	168
19	"	6	"	Switch doesn´t close electrical circuit	Use better and reliable switch	2	Use of ohmmeter to check electrical circuit	3	36
20	"	6	"	Spring of Tail Cap is broken	Use better and reliable spring	4	Control Spring	3	72
21	"	6	"	Spring of Tail cap is to short/weak	Use better and reliable spring	6	Control Spring	4	144

22	"	6	Intermittent energy supply from Energy Unit	Battery Case is worn out	Use better material for Battery Case	8	Use caliper to measure distance	2	96
23	"	6	"	Contacts of Battery Case are damaged	Use more durable material for contacts	8	Control contacts	2	96
24	"	6	LED Unit supplies light inconsistent-ly	Electrical contact of LED torn off	Use better and reliable LED	8	Use of ohmmeter to check electrical circuit	3	144
25	"	6	"	LED is weak connected	Use more solder	9	Control LED connection	3	162
26	"	6	"	LED Unit spring is broken	Use better and reliable spring	4	Control Spring	3	72
27	"	6	"	LED Unit spring is to short/weak	Use better and reliable spring	6	Control Spring	4	144
28	"	6	"	No contact between solder parts	Use better solder / alloy	5	Use of ohmmeter to check electrical circuit	2	60

29		6	"	Solder Joint has material failure	Use better solder and establish better soldering	10	Use of ohmmeter to check electrical circuit	1	60

Figure 8: FMEA Form

6. Pareto Analysis

An important tool to analyse and evaluate the FMEA form is to apply a Pareto Analysis to the data which is also known as ABC Analysis or Lorenz Distribution. Basically, Pareto Analysis visualizes the rank order of the influencing variables of relevance to one particular issue. By ranking the factors, in our case failure modes, according to their level of influence we want to get a structured overview on the FMEA results. Based on the Pareto Chart it is easy to identify the most important failure modes. Usually it is recommended to solve the failure modes with the highest impact first to enhance the overall product quality and reduce possible costs. In the FMEA sheet the RPN (Risk Priority Number) is calculated for each failure mode. The RPN value for each failure mode is calculated as a product of the importance, occurrence and detection of the fault. The factors are ranked by their RPN values starting with failure mode 6 which has the highest RPN of 378. Failure mode 6 describes an energy shortage of the 3 x 1.5 voltage AA-batteries which results in a not lighting flashlight. Furthermore, the ratios and the cumulative ratios are listed in the table below because they are required for the Pareto Analysis. For failure mode 6 as well the RPN-% value as the cumulative RPN-% value is 11.55%. According to the standardized Pareto Chart, the RPN-% values are visualized as bar graphs whereas the cumulative RPN-% values are shown as a curve. To be able to analyse the data in one chart two vertical axes are used that represent the RPN-% and the cumulative RPN-% values. On the horizontal axis the ranked failure modes are described. After the data has been classified and the mentioned values for RPN have been calculated, the Pareto Chart can be created to visualize the data. Generally, the Pareto Chart helps the team as well as the clients and employees from other departments to easily understand the influence of each failure mode. When looking at the Pareto Chart of the flashlight in Figure 10 it is obvious that the first two failure modes have a major impact compared to the others. With a cumulative RPN-% of nearly 20% the improvement or solving of those failure modes would mean a big improvement of the flashlight. Looking at failure modes 3 to 29 the RPN-% values slowly decrease without any big gaps in the process. To structure the Pareto Chart we can use 3 different intervals ranging over the following values:

1) *Interval A:* (6%,100%]   2) *Interval B:* (2%,6%]   3) *Interval C:* [0%,2%]

In the results shown below the statistical phenomenon of Pareto distribution can be found. This special distribution appears in many cases all over the place and basically means that in every set of items a very small number of high values has a much bigger impact on the result than the larger number of small values. In the underlying case only two failure modes, a relatively low number, appear very often and therefore are the origin of most troubles. One has to admit that independency of the investigated items is necessary. Therefore, the number of practically relevant applications is very small and also cannot be guaranteed in this case.

In practice it is often understood the way that with 20% of the available resources 80% of all problems can be solved. The principle is broadly used in time and project management as a rule of thumb to no put big efforts into small details with low impact.

No.	Sorted No.	RPN	%	Cumulative %
6	1	378	11.55	11.55
16	2	270	8.25	19.80
1	3	168	5.13	24.94
10	4	168	5.13	30.07
18	5	168	5.13	35.21
25	6	162	4.95	40.16
4	7	147	4.49	44.65
5	8	147	4.49	49.14
21	9	144	4.40	53.55
24	10	144	4.40	57.95
27	11	144	4.40	62.35
8	12	126	3.85	66.20
7	13	112	3.42	69.62
11	14	105	3.21	72.83
22	15	96	2.93	75.76
23	16	96	2.93	78.70
9	17	84	2.57	81.27

20	18	72	2.20	83.47
26	19	72	2.20	85.67
12	20	70	2.14	87.81
28	21	60	1.83	89.64
29	22	60	1.83	91.47
3	23	56	1.71	93.18
13	24	45	1.38	94.56
17	25	45	1.38	95.94
2	26	42	1.28	97.22
19	27	36	1.10	98.32
15	28	30	0.92	99.24
14	29	25	0.76	100.00

Figure 9: Pareto Table

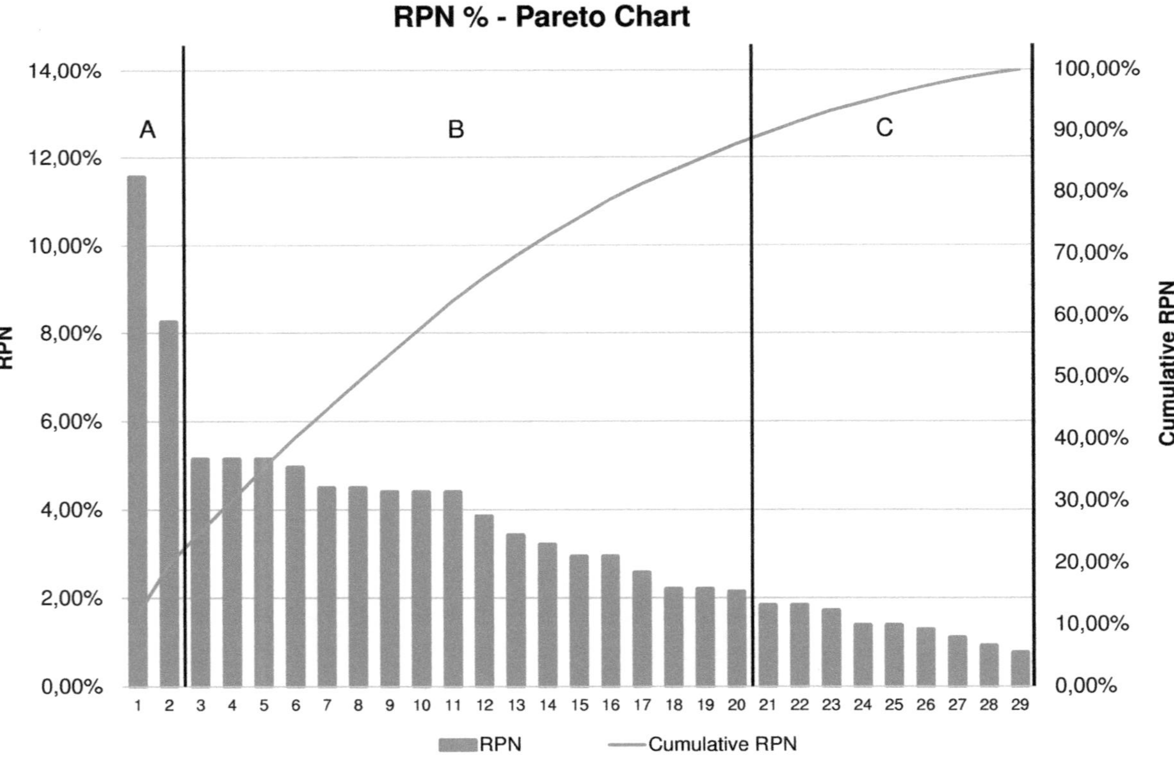

Figure 10: RPN % - Pareto Chart